BITCOIN IS REVEALED

How Builders, Markets, and Systems Shape Its Evolution

Jamil Hasan

Copyright © 2026 by Jamil Hasan

All rights reserved.

This book is for informational and educational purposes only and does not constitute financial, investment, legal, or tax advice. The views expressed are those of the author and are based on personal experience, research, and interviews. Readers should conduct their own research and consult appropriate professionals before making financial decisions.

This work reflects the author's interpretations and perspectives. While based on interviews and real-world experiences, any conclusions or representations are solely those of the author. Reference to any individuals, companies, or projects does not imply endorsement or affiliation unless explicitly stated.

Digital asset markets are highly volatile and speculative. Past performance and discussed scenarios are not indicative of future results.

Some names, stories, and identifying details may have been modified to respect privacy.

First Edition · 2026
Published by Crypto Hipster Publications

ISBN: 978-1-972991-02-2

Cover design by Jamil Hasan
Interior design by Jamil Hasan

All trademarks and registered trademarks are the property of their respective owners.

Printed in the United States of America

For my bernedoodle, Daisy—who never strays too
far from my side

Bitcoin is Revealed

You don't understand Bitcoin by defining it.

That's the first mistake most people make.

They look for a single explanation—something clean and stable that captures what Bitcoin is. Money. A store of value. A protocol. Infrastructure. A hedge. A network. Each of those ideas is correct. None of them is complete.

The longer you stay in the space, the more obvious that becomes.

Not because Bitcoin lacks structure, but because it operates across layers that don't collapse into one frame. Every perspective captures something real. The problem is assuming any of them captures the whole.

That assumption breaks quickly.

If you listen enough to people who are actually building in this space—not commenting on it, not trading around it, but building—you notice something uncomfortable.

They don't agree.

Not on the edges. On the fundamentals.

What Bitcoin is.
What it's for.
What matters.

And they're not guessing.

These are people operating inside the system—designing infrastructure, navigating markets, executing actual products, and defending core principles. Their disagreements aren't noise. They are signal.

That's where the shift begins.

The question stops being:

"What is Bitcoin?"

And becomes:

"Why do people who understand it deeply see it so differently?"

That question doesn't give you a definition.

It gives you a framework.

Bitcoin is not something you define once and move on from.

It is something that reveals itself through interaction.

Through the systems built around it.
Through the pressure applied to it.
Through the realities that filter what survives.
Through the resistance that protects what matters.

This book is built around those forces.
Not to simplify Bitcoin—but to make it legible.

You'll see Bitcoin through four lenses:

Structure and Access — how it enters the real world
Pressure and Behavior — how it responds under stress
Execution and Reality — what actually works
Preservation and Drift — what risks being lost

Each lens is incomplete on its own.
Together, they don't give you a definition.

They give you something better:
A way to see what's actually happening.

This is not a book about price.
It's not a book about cycles.
It's not a book about prediction.

It's a book about how a system evolves when no one
is in control, but everyone is participating.

If you're looking for certainty, you won't find it
here.

If you're looking to understand how Bitcoin is being
shaped—in real time, by real people, under real
constraints:

Then keep reading.

Bitcoin is Revealed

Preface

I'm Jamil Hasan. People call me the Crypto Hipster—a name I gave myself, and one I've chosen to own. It started as a joke. A meme. Then it became a signal. Not rebellion for its own sake, but a way of seeing things.

Crypto moves quickly—often faster than understanding can keep up. Sometimes, a topic I spent several podcast episodes fully understanding in the first week completely changed the next week.

There is a tremendous amount of noise in the market and not enough uncovering of the true signals. I aim to change that.

Discussion about the price of Bitcoin, meme-coin hype, 1000X speculation, and narrative repetition never truly interested me. I wanted to find out what genuinely matters.

So, I started the Crypto Hipster Podcast in February 2021. Not to chase headlines. To have actual conversations with people building this space.

Founders. Creators. Thinkers...people trying to reshape systems.

Crypto Hipster became more than a podcast. Nine seasons. Over 580 conversations. Hundreds of written works.

Then I hit an uncomfortable fact:

Curating discussions is not the same as creating ideas.

For years, I documented conversations. Published transcripts of all my podcasts. Built an archive. A colleague told me that if I wanted to be an influential leader, I would have to do more than simply host conversations. I would have to move the conversations forward.

That person was right.

This book is the first of sixty-seven Crypto Hipster Curtain Calls compilations spanning two-hundred eighty-one interviewees, is the first part of that shift.

Curtain Calls differ significantly from my first 399 books. It is not a collection of transcripts. It's an interpretation. A reflection. A synthesis of conversations, ideas, and lived experiences. It's where my voice meets the voices of the people I've learned from.

Crypto is not just technology. It's about power. About identity. About freedom. And above all, about people.

My path into this space came not from comfort. For the past five years, I have fought an aggressive desmoid tumor. What began as a way to keep my mind occupied grew into something larger. It became a way to process the world and my place in it.

This book blends two things:

- The conversations I've had with builders of the digital economy.
- And the lessons I have lived myself.

Conversations with others. And personal memoirs interwoven with them.

You'll see both here. You'll see ideas about decentralization and finance. About infrastructure and governance.

You'll also see stories. Resilience, failure, faith, relationships, recovery.

This technology does not exist in isolation. It reflects who we are. Too often, crypto gets reduced to price charts. The authentic story runs deeper:

- Whether these systems actually change anything or just recreate the same structures with a new label.
- Whether innovation leads to freedom or just a different method of control.
- Whether we, as individuals, choose to take part on purpose or just go with the current.

This book is my attempt to slow that down. To take hundreds of conversations and extract what matters most.

To connect patterns.

To ask better questions.

To share what I've learned. Not as answers. As a framework for thinking.

This is not only my story. It's an invitation...an invitation to think differently. To question more. To build with intention.

I hope that if you see yourself in these pages, this book will serve as an inspiration for you to write your own story with more clarity, purpose, and conviction.

Table of Contents

The Search for Definition

I used to think the goal was to understand Bitcoin—
not through its price, the cycles, or even the
technology itself.

I thought there was something steadier under all of
that. If I listened long enough and asked the right
people the right questions, I would land on a
definition that worked everywhere. A single idea to
anchor everything else. It faded slowly,
conversation by conversation, perspective by
perspective.

What I found in the end wasn't the clarity I'd
imagined. It was harder to accept. There is no single
understanding of Bitcoin. There are only
perspectives. It sounds obvious, but rarely is.

Most people treat Bitcoin as if it will finally resolve
into one thing. Money, a protocol, a store of value, a
system of freedom—or simply speculation. Each

view carries conviction, can be justified with logic and evidence, and feels complete to the person holding it. But together they don't merge. They sit side by side and don't add up.

That was my first clue that something was off, and the second was who disagreed. Not casual observers. Not people on the fringe. The disagreement came from people deep inside the space. Builders and founders, developers and operators—people who had spent years working in the system.

They weren't arguing about unimportant details but about fundamentals. What Bitcoin is. What it is for. What matters. And they were precise about it.

That shift changes how you listen. If disagreement comes from missing information, you can fix it. Update your model. Get better data. You converge— but when disagreement comes from different starting points—different incentives, different goals, and different beliefs—convergence stops being the default. Divergence takes over.

So the question shifts—from "What is Bitcoin?" to "Why do smart, experienced people see the same system so differently?" That question doesn't give you a definition. It gives you something to watch.

I stopped trying to pull one truth out of every conversation and focused on the lens people used. Not just what they said, but how they got there. What they emphasized. What they brushed off. What they treated as obvious.

Those signals were consistent. Not across everyone, but within each person. Each person had a frame, and that frame shaped everything they saw.

It didn't hit me in a single moment but built up over time. Repeated exposure to interpretations that didn't just differ but often contradicted each other. One person described Bitcoin as a financial instrument. Another talked about it as a protocol. Someone else saw it as a political construct. Another treated it mainly as a tool.

When I first encountered Bitcoin, I saw it as a path to freedom. Over the past nine years, that view has shifted—from what it could have been to what the overwhelming majority now believe it is.

The system didn't change; the realities people built around it did. At first, I tried to reconcile these views. Find the overlap. Reduce differences. That instinct comes from training. We're taught to simplify complexity, to pull systems into essential parts, and to collapse variation into one story.

That approach works in many places, but not here. Bitcoin resists reduction not because it's misunderstood but because it operates on multiple levels at once. Technical, economic, political, and behavioral layers. Each gives a valid but incomplete view.

No single layer holds the whole definition, and isolating one means losing something. Seeing that changes the goal. You stop trying to define Bitcoin in a vacuum. You watch how people define it in practice. Through what they build. How they use it. What they prioritize.

This shift didn't feel like a choice. It just happened. The noise didn't disappear. It lost relevance. What remained was signal. Not the loudest voices. Not the most repeated narratives. The perspectives that held up under pressure.

Once I listened differently, patterns became obvious. Some conversations circled the same surface-level ideas. Others, quieter at first, revealed deeper structures. Not in words, but in how people related to the system.

That's where the actual difference shows up. Not right versus wrong, but orientation. Some aim to fold Bitcoin into existing systems. They build pathways to make participation easier. Others focus on external forces—liquidity, instability, global stress—and how those shape its behavior. Some engage by testing what works. Execution over theory.

And then there are the preservationists. They worry less about growth or adoption and more about keeping certain things unchanged. That last group is often misunderstood, as their view doesn't always align with growth or ease of use. But it lines up with the original purpose of the system.

Then a pattern shows up. Bitcoin doesn't present the same way to everyone. It reflects the priorities of whoever is using it—their values, their goals, what they care about. Bitcoin itself doesn't change because of these perspectives. What changes is what becomes visible.

This reframing changes how you approach the system. Instead of asking which interpretation is right, ask what each interpretation reveals. What it brings into focus. What it hides. And how those choices, when turned into action, shape the system.

These are not passive interpretations; they lead to decisions. Builders don't just describe systems. They shape them. By what they prioritize, what they ignore, what they simplify, and what they preserve, they influence how Bitcoin is experienced.

The effects aren't always immediate; they accumulate. Over time, they create direction. This is how Bitcoin grows. Not by centralized design but through distributed influence. No one controls it, yet everyone contributes.

That's easy to miss because Bitcoin feels fixed—its supply is fixed and its protocol is stable. The base layer moves slowly, if at all. But everything around it is moving. How people access it, use it, and understand it.

For most people, those surrounding layers are Bitcoin. They don't touch the protocol. They interact with systems built on top, and those systems shape their perceptions. Perception matters because it has real consequences.

Access through custodial platforms makes it function as an asset. Held in self-custody, it becomes an instrument of sovereignty. Integrated into applications, it looks like infrastructure.

The underlying system remains the same. The experience—and the meaning—shifts. This is already happening, and it is accelerating.

The more concrete question becomes: what kind of Bitcoin is being built? Not in theory, but in practice.

There is no single answer. Multiple versions exist simultaneously. Sometimes they align. Sometimes they clash. They are constantly changing.

If Bitcoin were simple, this wouldn't happen. If it were only money, debates would converge on monetary properties. If it were only tech, it would standardize on implementation. If it were only ideology, people would split into clear camps. Instead, Bitcoin sits in all those places. That creates friction. Especially for people who want to reduce systems to one explanation.

Remove that expectation, and a different clarity appears. Not by simplifying, but by seeing structure. A developer sees constraints and tradeoffs. An institutional allocator sees exposure and risk. A founder sees an opportunity. A trader sees volatility. A policymaker sees disruption. Each view is incomplete, but together they create a fuller picture.

The aim isn't to unify them, but to understand how they interact and what follows. Those interactions play out in markets, in narratives, and in infrastructure. Sometimes quietly. Sometimes all at once. When they intersect, something gets selected. Not by agreement, but by implementation. That drives change.

Bitcoin evolves because people act—something is built, adopted, or ignored. Those choices accumulate and move the system. This is where signals become concrete. Signal isn't what peaks in the moment. It's what lasts. It is what holds up across conditions and eras.

Crypto is noisy. New narratives, new tokens, and new promises appear constantly. They grab attention, then fade. Bitcoin works differently. It doesn't need novelty. It persists across cycles and shifts in attention. That persistence gives it weight.

It also makes interpretation harder. Bitcoin stays relatively stable while everything around it changes. Markets shift. Technology improves. Different people come and go. Each cycle brings new expectations, and Bitcoin absorbs them while keeping its core. That creates tension.

The system is stable, but the demands on it are not. Some push for faster scaling. Some resist change entirely. Some want integration. Some want independence. These tensions aren't failures. They are constraints. And constraints force choices. Those choices shape outcomes.

Builders are central not because they control Bitcoin, but because they act within it. They don't wait for everyone to agree. They pick a direction and build. Implementation eventually matters more than opinion. A product launches. A standard is adopted. A system integrates. Advocates appear on the Crypto Hipster Podcast.

Each move nudges evolution—not through drama, but through accumulation. Bitcoin rarely shifts in obvious ways, but it is constantly being shaped at the edges—where most people encounter it.

And that's where real evolution happens.

The System Reflects

People talk about Bitcoin in many ways—money, protocol, asset, network, ideology. All of it is true, and all of it is incomplete.

The problem is not a misunderstanding. The problem is stopping at the first explanation that fits. Once a frame works, people stick to it, and over time, everything reinforces that first view.

A macro investor treats Bitcoin as a hedge against systemic risk. A developer thinks in terms of constraints and tradeoffs. A founder sees a chance to build. An ideologue sees a statement about how the world should work. Each view makes sense on its own, pointing at something real while leaving important aspects out.

That omission is usually structural, not deliberate, because no single view can hold the entire system. Bitcoin operates across layers that do not collapse into one another, which is why thinking of it as a mirror helps. Not a neat metaphor, but a practical description of how it behaves.

A mirror does not make an image; it reflects what stands in front of it. What you see depends on where you stand and what surrounds you. Bitcoin works the same way. It does not force a single meaning; it reflects the orientation of whoever is engaging with it. That orientation changes through experience, incentives, and exposure to other systems. Once it forms, you stop looking beyond it.

Two people can study Bitcoin for years and end up with conclusions that clash. They are not looking at different systems. They see different reflections of the same system, shaped by the lenses they bring. A lot of confusion comes from assuming disagreement means someone is wrong.

In many areas, that is true. Here it breaks down. Disagreement often just shows the system's structure. Multiple interpretations can be valid until they stop aligning.

Start with incentives because they tell you what matters. A trader focused on short-term moves sees volatility as an opportunity. A long-term holder reads the same volatility as noise. An institution calls it risky. This is the same event with different meanings because the goals differ.

Constraints work the same way. They set what's possible and what's practical. A protocol developer thinks about block size, throughput, and security tradeoffs. A user on an app rarely meets those limits. Their view is shaped by layers that hide complexity and simplify it. That shaping decides what is visible and what is hidden.

Those layers are not neutral. They encode choices about what to stress and what to ignore. Those choices shape experience, which shapes understanding.

At that point, the mirror becomes recursive. What people see affects what they build. What they build changes what others see. A feedback loop forms. No single actor controls it, but it has direction. Some interpretations get implemented and adopted, and as they do, they gain momentum. Not because they are truer, but because they are made real.

This is how systems evolve without a central plan. Not by unanimous agreement. By accumulation. That matters because it shows the limits of pure theory, which can argue indefinitely about what Bitcoin should be. What matters is what it becomes in practice. And practice is shaped by builders dealing with constraints.

Builders do not work in the abstract; they work where technical limits, regulations, and markets meet. Those constraints force choices, and those choices lead to outcomes. Those outcomes shape how people experience the system. That is why the same protocol can yield very different realities.

Think about access. One path is self-custody, which includes private keys, direct control, and full responsibility. Another is custodial platforms, the more familiar world of accounts, interfaces, and managed risk. Both touch Bitcoin, but they feel different. One is direct participation. The other is exposure.

This is not just a philosophical split; it is behavioral. People with direct control act differently than those behind intermediaries. They take different risks, make different decisions, and care about different things. Those differences compound.

The same thing happens with infrastructure. Layer twos, side-chains, and apps are built to fix scaling, speed, privacy, and usability. Each solution brings tradeoffs and shows priorities. As they grow, they change how Bitcoin is used and understood.
This is where the central tension emerges. Bitcoin's base layer stays mostly stable as everything around it keeps changing. Most systems change at the core,

but Bitcoin changes at the edges. Edge-driven evolution lets multiple meanings coexist, but also fragments perception. Users encounter different versions depending on how they access it and what has been built for them, with builders sitting at the center of that process. Builders sit at the center of this—not as users, but as translators. They take an open, abstract system and make it usable in specific contexts.

In the process, they create meaning. Not by forcing it, but by shaping experience. That shaping involves tradeoffs, such as deciding what to prioritize, what to optimize, and what to sacrifice. These choices are often invisible to end-users, but they define how the system is encountered. And once an experience is set, it steers interpretation.

The mirror does more than reflect; it influences what will be reflected next. Not by control. By direction.

Use changes what others see, and patterns form—without fixed outcomes or dominant tendencies. Bitcoin is an asset, infrastructure, protection, and an opportunity. These views coexist. Their weight shifts with markets, tech, rules, and who joins in. Those forces sit outside Bitcoin's base layer but shape how it is used, with that use defining what the system becomes.

Neutrality gets complicated. At the protocol level, Bitcoin is neutral as it enforces rules without choosing sides. But neutrality does not follow into use. Use shows intent, and because intent varies, it creates a layered system. Multiple paths develop at once that sometimes align and sometimes clash.

Trying to pin Bitcoin down often fails because people isolate one layer and ignore the interactions that actually drive change. Time matters a lot. Short-term behavior misleads as stories shift fast, attention moves, and new takes appear. Deeper patterns take time. They need repeated choices, steady implementation, and sustained use. That is when the signal separates from the noise.

Seen this way, Bitcoin is less a fixed thing and more a process that keeps getting shaped by people who use it, build with it, adapt it, and sometimes protect it. They are not trying to define it; they are trying to make it work. Each action pushes the system. Unevenly, but meaningfully. This idea ties into what follows.

The four builders in this book do not stand for the entire system. They are not exhaustive. They were picked because each shows a different way Bitcoin gets shaped. Each has a different orientation, accepts different tradeoffs, and builds toward a different result. Alone, their views can feel incomplete or even at odds; together they reveal structure. Not by settling differences. By showing them.

Before we look at those views, note that this dynamic unfolds in stages. Early on, ideology ruled. The system was small and participation was limited. Many people shared similar beliefs.

Bitcoin was a statement, a pushback against existing systems. The language was freedom, sovereignty, and decentralization. These were reasons people joined, and not just features.

Then growth changed that. New actors came for other reasons: financial, technical, and strategic. The shared story split, and Bitcoin's original frame no longer held everything together. Multiple interpretations grew in parallel, and each projected different goals onto the same protocol. That is where Bitcoin has been for some time.

Now the mirror reflects many orientations at once. Tension shows up. Early adopters versus new entrants. Ideals versus practical needs. Simplicity against scale. These tensions are structural. They appear whenever a system grows beyond its start.

Most systems react by converging to a dominant model, reducing complexity, and collapsing variation into a standard form. Bitcoin resists that. Not by design, but by how it evolves. With no central plan, multiple interpretations can stick around, offering a persistence that breeds uncertainty. With no single path ahead, it is filled with competing directions. They do not get settled by debate; they get settled by building.

Back to the builders then. Not lone actors, but as a group, builders turn ideas into structures, products, extensions, access points, and interfaces. Each choice says what Bitcoin is for in practice. You do not need consensus to ship, but you need execution. Once something exists, it shapes behavior. Users see what has been built for them and not the full system. What exists becomes what is used. What is used becomes what people expect.

Those expectations shape the next build, and that recursive process is how influence works without central control. Not through authority but through availability, every additional layer adds a bias toward certain uses, behaviors, and interpretations. These biases pile up. Not equally, but with direction.

That is why protocol neutrality is hard to maintain in practice. User-experience is molded by surrounding systems; and those systems reflect builders' priorities, which do not match. Some optimize for speed. Some for security. Some for compliance, privacy, or profit. Each brings tradeoffs that shape the system more than any single feature.

Real-world constraints reinforce this. Technical limits are only part of the equation. Regulation, economics, and human behavior all shape what can be built and what gets prioritized. A system designed for compliance looks different from one designed for sovereignty. A setup optimized for institutional capital differs from one built for individual control.

These are not abstract differences. They create distinct structures and distinct experiences. Divergence shows up in outcomes, not in ideas, but in what products exist, what features get attention, and what tradeoffs are accepted. Those are the signals that define the system.

Understanding Bitcoin means moving past a single definition. Watch how forces shape different versions of the system and how those versions interact. They do not act in isolation; they have shared infrastructure, market dynamics, and linked narratives. A product built for institutions affects liquidity. Liquidity affects price. Price affects perception. Perception affects participation. The system is connected, and that connection makes linear cause and effect a poor model.

Feedback loops dominate. Small decisions can matter a lot over time through accumulation. Patterns come from trajectories about where the system is heading and why. Then the observer's job changes and becomes less about parsing single views and more about tracking how those views together push the system. That needs different attention: less focus on statements and more focus on outcomes.

Clarity comes from mapping what gets built, what gains traction, and what is abandoned—not from simplifying the system.

Seeing how parts interact and produce direction is the framework of this book. It is not a fight over definitions; it is an exploration of how builders engage with Bitcoin, what they prioritize, what they ignore, and what they shape. Each builder is a lens. Through them, the system becomes clearer. Not whole, but structural. And the structure is enough to see what might come next.

Before we jump into those perspectives, remember these lenses are not abstract. They are people making real decisions under real constraints.

Sung Choi, senior vice president of business development and operations at Coinme, with whom I spoke on January 17, 2025, works at the meeting point of Bitcoin and institutional systems. He does not argue about the meaning of Bitcoin; he is making Bitcoin usable where clarity, compliance, and accountability matter. Through his work, Bitcoin becomes integrated and accessible, but that integration sets boundaries that change the experience.

Agne Linge, Head of Growth at WeFi, on June 1, 2025, explained how she looks at Bitcoin from a macro view. She watches how global markets, policy shifts, and capital flows affect it. Agne treats Bitcoin as a system that responds to changing conditions, and not as a fixed object.

Mark Højgaard, co-founder of Coinify, shared with me on March 24, 2024, how he focuses on execution. He cares about what works in payments, infrastructure, and user systems. For him, Bitcoin is filtered by real constraints and discusses how only what runs reliably sticks.

Vlad Costea, founder of the Bitcoin Takeover Podcast (March 16, 2024), is a counterweight. Where others stress integration and function, he stresses preservation. Vlad aims to keep core principles intact as things change. His view adds tension, not about what works, but about what should stay.

Together, these perspectives do not define Bitcoin. They reveal it.

Structure and Access

There's a version of Bitcoin that lives only in theory—coherent, elegant, and internally consistent. In that version, the system works as designed, where it is decentralized, permissionless, and hard to control.

Participation is direct, ownership is absolute, and rules are enforced by code rather than institutions, giving it a self-contained quality where there is nothing to add or negotiate. It just exists. That version is authentic, but incomplete.

Once Bitcoin leaves that closed loop—beyond people who get it and beyond those who'll use it on its own terms—it runs into another reality. That reality does not run on elegance alone; it runs on requirements, does not reward purity, demands structure, and does not adapt to the system. The system must adapt to it.

This is the world Sung Choi works in. "We're not trying to explain Bitcoin—we're trying to make it usable where it otherwise wouldn't be." He doesn't start with what Bitcoin means as an idea. He starts with what Bitcoin must meet to work inside systems that already exist. These are real operational systems that don't allow ambiguity, not because they lack imagination, but because they carry responsibility—legal responsibility, financial responsibility, and reputational responsibility.

In those systems, uncertainty isn't an intellectual puzzle. It's a risk to manage, and that risk changes everything. Seen that way, the question is no longer whether Bitcoin works in principle, but whether it can be made legible to systems that need clarity before they join in. That clarity depends on whether it can be accessed, measured, and controlled, at least at the interface by entities that need those assurances.

This isn't philosophy; it's practical. At Coinme, Sung's job isn't to lecture about Bitcoin, but to make it usable inside places that otherwise can't touch it. He builds infrastructure to turn an open, permissionless protocol into something that can run inside closed, regulated frameworks. This is not accomplished by changing the protocol, but by shaping how people access it. That is a small but crucial distinction, because Bitcoin does not change—how people experience it does. And experience comes from the surrounding systems.

When Sung talks about Coinme's work—APIs that let partners add crypto features, compliance setups that meet regulators, licensing that lets operations across borders—he's not describing Bitcoin itself. He's describing the conditions that let people outside the core participate. Those conditions are not optional; they are prerequisites.

Scale needs more than just possibility; it needs reliability and predictability. A structure that lets people decide without taking on too much risk is where tension shows up. Bitcoin's original form minimizes trust, removes intermediaries, and lets people deal directly with the system with no permission needed.

But the systems Sung deals with are built differently. They need intermediaries, oversight, and ways to be held accountable. These aren't ideological choices; rather, they're structural necessities. A bank can't put money into something it can't model. A partner won't integrate a service it can't count on. A regulator won't allow access to a system that doesn't meet set standards for security and compliance. "Institutions don't operate on possibility—they operate on what they can defend."

Those constraints don't disappear when Bitcoin is involved. They pull into it; and when they do, the way people access Bitcoin changes at the edges. Most people interact at those edges, and very few run nodes, hold private keys, or build transactions by hand. Instead, they use apps and platforms that hide the messy parts. "Most people will never touch the protocol—they'll use what's built around it."

Those abstractions aren't neutral. They encode choices about custody, about identity, about how risk is handled and access is given. When Sung talks about a "native experience" that feels seamless—no jumping between systems, no need to understand the mechanics—he means removing friction to make it familiar and match expectations set by existing financial and tech systems. That's usability.

People often treat usability as something to fix later, once the core works; but in places where participation is broad, usability matters from the start. People don't adopt things they don't understand unless the things are made accessible in ways that don't require understanding. That's the paradox. The more complex the core is, the more you must simplify the surface. Simplifying means abstracting.

When you hide complexity, control shifts partially, at least enough to change how people interact. Someone who holds their keys acts differently than someone on a custodial platform. The first takes direct responsibility, while the second relies on intermediaries. Both are real, valid, and different.

Sung's work becomes interpretive, not just technical. Every choice about how to give access reflects what matters most: control, convenience, compliance, and scale. These priorities don't line up cleanly; they force tradeoffs that shape the system more than any single feature.

The regulation shows this clearly. When Sung explains how existing financial plumbing built for money transmitters gets extended to crypto, he's pointing out that Bitcoin doesn't float alone. It intersects with systems that already have rules that are enforced by institutions, not by code, and brings added costs.

Compliance programs need resources. Licensing takes time, and identity checks add friction. These safeguards are not wasteful steps, but they slow things down. Adding intermediaries changes how people interact. For people who want speed and directness, this can feel like moving backward; but for those who need stability and oversight, it's the only way in.

Sung navigates both sides not by picking one, but by working where both must be served. That space isn't clean or consistent with one ideology, but it's where large-scale participation happens. Payments show this in practice. Bitcoin was pitched as peer-to-peer money: fast, borderless, and efficient. That wasn't wrong, but it was incomplete. Real payment systems add other factors: volatility, which affects prices; fees, which change cost predictability; and settlement times, which shape user experience. These operational limits are not abstract and must be accounted for.

Stablecoins show up as a complement, not because they match Bitcoin's original idea, but because they solve issues in some contexts. They stabilize price, smooth transactions, and fit existing expectations for payments better.

Sung doesn't see a contradiction. He sees functionality in using different tools for different jobs, with Bitcoin as a base and stablecoins as a transaction layer. Infrastructure is about tying the base and different layers together. This is how complex systems evolve: not by replacing everything, but by adding and adapting pieces. Over time, you get something messier but more useful. That mess doesn't always match early hopes, but it matches how systems work under constraints, which then forces choices.

What must work? What can be simplified? What can wait? Those questions steer development more than abstract ideals. Sung lives by those questions not as theory but as daily reality. What will partners actually use? What can integrate without unacceptable risk? What can scale past early adopters?

You cannot answer those questions with theory alone; they require testing, iteration, and adjustment. So, Bitcoin as a static thing blurs. While the protocol remains stable, the user experience keeps changing. Builders like Sung drive that change, which happens not through central control but through many small choices about access and use. Each choice may seem insignificant, but together they set a direction. That direction isn't set in stone, as it is shaped by constraints, incentives, and priorities.

This is the Bitcoin that comes out of Sung's view. Not the original design. Bitcoin being translated into places that need structure doesn't alter Bitcoin's core, even though it alters how people experience it. And for most users, experience is reality.

As Bitcoin becomes more structured, it becomes easier to use, gets more integrated, and looks like the systems it once stood apart from. Not totally, but enough to ask when integration changes the system.

Sung doesn't answer that. He works within the question. The alternative would keep Bitcoin perfectly pure and limit its reach. The major trade-offs become purity versus scale, directness versus accessibility, and independence versus integration. There is no neat, simple resolution; there must be a shifting balance that moves as new people join, new constraints appear, and new structures get built.

This is Bitcoin at scale: not fixed, but shaped at the edges by people making it usable where possibility is not enough and where structure is required. And structure does more than allow participation. It defines it. "If you want scale, you have to accept structure."

There comes a point when systems stop being judged on potential and start being judged on consequences. That point sneaks up through exposure, through repetition, and through added risk. At that point, who enters changes. Early on, people accept uncertainty, experiment, and live with volatility while they learn. But as systems grow, tolerance for risk shrinks, and people become less bold because the cost of failure grows. That's especially true for institutions.

An individual can try something new and take a hit. An institution can't. Institutions sit inside layers of accountability, which are internal, external, regulatory, and reputational. Each layer sets limits, which change what people do. Decisions are rarely made just on what's possible; they're made on what's defensible, justifiable, explainable, and survivable.

This is where Sung's view matters. He's not in a place where curiosity is enough; he's in a place where choices need to be defensible to partners, regulators, and stakeholders. That flips the problem at the engagement level instead of at the design level. Seen from there, Bitcoin isn't just an opportunity. It's a variable inside a bigger system that needs to be judged for its effect on what's already running.

How does it change risk exposure? How does it fit with compliance rules? How does it touch day-to-day operations? Those questions don't come from Bitcoin itself. They come from the systems Bitcoin runs into.

When those systems engage, they shape the interaction, which is how integration happens. Integration occurs not by everyone's agreement, but by accommodation. Bitcoin doesn't bend to those rules, but the access layer does, and as a result, that layer becomes the control point. "Access is the real control layer." It is not the control of the protocol, but the control of participation that determines who gets in, how they get in, and under what conditions.

Infrastructure begins to resemble governance, containing not only rules but pathways that define what is possible, what's allowed, and what's nudged. When infrastructure draws those paths, it shapes the system for most users. You don't always see it, but you experience it as convenience: smooth onboarding, integrated flows, and familiar screens. Under that convenience are choices about identity, custody, and compliance. Each one nudges the system, often in ways you don't notice right away.

That's why "invisible crypto" matters. Users want the difficulties to vanish so transactions feel instant and interfaces look like other apps in a system that behaves like what they already know. That's what Sung is building toward. Not because it matches Bitcoin's original vibe, but because it matches how people actually adopt new tools and technology.

People don't enjoy changing habits. They adopt things that fit what they already do. Changing habits creates tension. The more Bitcoin is folded into familiar tools, the less it stands out, the less it asks people to be active, and the less it asks them to learn. Engagement slides toward reliance. People rely on the system, on the interface, on the provider, and on the process. That trust isn't wrong, but it's different from the trust model Bitcoin started with.

Bitcoin was meant to reduce how much you had to trust others. Structured systems bring some of that trust back, though not all of it. Sung's work sits in that space. He's reconciling models of trust, which include trust in code, trust in institutions, and trust in processes.

Each trust model has pros and cons, and they get mixed. They do not fit perfectly together, but because they're all needed, the result is hybrid systems that are partly decentralized and partly structured. These hybrids do not fit any neat story—they are neither fully independent nor fully controlled, but sit in the middle. Most real-world activity happens there; and that's where the mess lives.

Hybrid systems are hard to explain. You must juggle technical, regulatory, and operational layers, and how those layers interact. Although users rarely see that complexity, it doesn't vanish. It is managed by infrastructure and by companies like Coinme.

Managing complexity brings influence, and people who manage the hard parts shape outcomes. Their choices matter, not because they set out to, but because they determine what features get built, what risks get cut, and what experiences get smoothed. Those choices aren't neutral. They show what matters, and over time, these choices create patterns.

Some use cases get pushed forward while others drift away. Those that drift are not wrong, but they're just not supported. Systems do not change by wiping out alternatives; they change by making certain paths easier.

Sung's role isn't to pick the final shape of Bitcoin, but it is to open ways for specific participation that fit structured settings. That brings scale and standardization. Standardization makes things simpler, but it narrows variety because another tension shows up—flexibility versus consistency.

Bitcoin in raw form is flexible, and there are lots of ways to use it. Structured systems take some of that flexibility away as they trade it for predictability. Predictability is needed if you want to plug into other systems, but it costs options. Most people accept that trade-off. Predictability matters more when the stakes are high, although some see it as a loss—not of features, but of possibilities.

The long-term effect is worth thinking about. As more people enter through structured channels, their idea of Bitcoin changes because their experience shapes expectations. Those expectations shape what builders make, and builders follow demand, which follows experience. It's a circular loop. Structured access makes for structured expectations, and structured expectations steer development.

Over time, that shifts the center of gravity. Not away from Bitcoin's core, but toward certain ways of seeing it. That's not inevitable. But it's directional. And direction matters because it shapes what the system becomes in use, not just in theory. Through Sung's eyes, Bitcoin looks more accessible, more integrated, and more structured. So, it ends up being shaped by the environments it enters.

That doesn't erase its basic properties, but it changes how people experience them. For most users, Bitcoin turns into something they access, and not something they run; something they use, and not something they fully dig into. That difference defines the system at scale because it's how most people will encounter it.

Here's a key point. Structure isn't some add-on. It's a filter. It decides what you see and what you don't, what gets highlighted, and what gets hidden. That filtering shapes perception, and perception shapes reality. How Bitcoin is shaped at the edges will end up defining it for most people, not at the code level, but at the experience level. And experience, over time, becomes the definition.

People think structure is an external addition on top of technology, but at scale it stops being external and becomes part of how the system expresses itself. That expression happens not in the protocol, but in the places where most people interact. That's easy to miss because the core rules stay the same, with its supply unchanged.

With the rules enforced, the system still runs as designed. But from a user's point of view, everything can change as access is mediated, experience is shaped, and choices are narrowed substantially enough to set the default path. And default paths matter more than edge cases in big systems.

Most people do what's available, not what's possible. That's where Sung's impact is largest. Not by remaking Bitcoin, but by shaping how people meet it at scale. That meeting is filtered and rarely direct, through interfaces, processes, and rules that sit outside the protocol. Those filters are useful because they reduce friction, let people participate, and make otherwise incompatible systems work together. But they also add interpretation, which shifts the meaning.

This isn't unique to Bitcoin. All systems build layers as they scale, including finance, technology, and social systems. Abstraction lets you grow but creates distance from the mechanics and direct engagement. That distance changes behavior. Someone who uses the system directly will see different risks than someone who uses it through layers. They will notice other opportunities and value things differently. Those differences add up, causing different versions of the same system.

Sung's lens makes that clearer, not by saying one version is better, but by showing how one version wins in structured settings. The version that's easy to access, easy to integrate, and easy to rely on. Reliability is underrated in adoption. It doesn't make for exciting headlines or fuel early hype, but it decides long-term use.

Unreliable systems get explored while reliable systems get used. That's the line between experiment and adoption. Sung works on the adoption side. The goal isn't to prove something works; it's to keep it working consistently at scale.

Keeping it consistent requires standards because these standards cut variation and raise predictability. Predictability lets you integrate, and integration lets you scale. That path is how things move from niche to infrastructure.

Seen through Sung's eyes, Bitcoin gains more structure, becomes easier to reach, and becomes more tied into systems that need reliability and accountability. The core properties stay the same, but people experience them differently. Most users don't run Bitcoin themselves anymore. They access it through platforms, interfaces, and systems that try to cut friction and manage risk.

That shift changes how people act. Intermediaries appear; interaction gets simpler; and Bitcoin lines up with expectations from existing finance. The protocol itself hasn't changed. The rules remain, the supply stays fixed, and the base layer keeps working as designed. But what people meet is not the base layer; it is the layer built around it. And that layer shapes how the system is understood in practice.

You can see the tension there. The more Bitcoin is set up for access, the more it feels like the systems it was supposed to differ from. That is not because the design flips, but because the experience does. Participation gets easier and more mediated, making risk look smaller to an individual. But that risk spreads into other layers and brings new dependencies. Some call that progress because it lets Bitcoin scale, brings more people in, and makes Bitcoin usable where it otherwise wouldn't be. Others see it as a move away from the original intent.

Bitcoin becomes less about direct control and more about the surrounding systems. Both views make sense and are happening at once. That's what Sung shows. Not a redefinition of Bitcoin—a translation that makes the system legible to institutions, partners, and users who need structure before they join. The translation doesn't change what Bitcoin is, but it changes how people encounter it. And over time, that changes how they see it.

Perception matters. It shapes who takes part; and who takes part shapes where Bitcoin goes. As structured access spreads, it changes expectations. People use what is easy, and what gets used becomes normal. What is normal directs what gets built next. This is how influence works without central authority—simply through implementation.

Sung sits inside that process, not controlling the protocol, but shaping the paths people rely on to enter it. Those paths matter because they define most users' experiences, which become their idea of Bitcoin. Feedback loops matter. Builders follow demand; demand follows experience; experience follows structure. The loop continues.

Bitcoin doesn't settle into one shape. It grows in layers, each with its own priorities and tradeoffs.

Sung makes those layers visible as real operational processes, not only ideas. Structure lets people join and decides what joining looks like. And that keeps changing, slowly and mostly at the edges, where choices about access, usability, and integration add up. That is where this version of Bitcoin takes form. Not down at the protocol. But at the interface between the system and the user.

And the next pressure acts there.

Pressure and Behavior

There's a real difference between knowing a system on its own and knowing it when things go wrong. Alone, systems look clean because you can see the structure, follow the logic, and predict behavior up to a point. Under stress, that neatness cracks as otherwise stable links expose hidden dependencies, assumptions stop working, pieces that seemed separate act as one, and pressure in one area transmits to another.

Agne Linge doesn't begin with Bitcoin or treat crypto as its own world, but starts with finance as a network that is fragile and shaped by forces outside any single market. Her background is in traditional finance. She saw the fallout from the 2008–2009 crisis up close, and she spent almost a decade in crypto. That mix makes her less of a cheerleader and more an observer of how systems respond under pressure. "You don't really understand a system until you see how it behaves when it's under stress."

Once you have seen a system fail in real terms, your habits change, and you stop assuming stability. Risk isn't a knob you can turn because it builds up quietly and then it bursts. Crises don't just break systems; they break the links that made the system seem strong.

Agne pushes back on treating Bitcoin as if it exists in isolation, as that approach is not useful. Bitcoin must be understood within the wider picture of global finance, political decisions, capital flows, and macro forces. Agne says everything is linked, as crypto connects to risky assets and spills into the wider economy. Trying to separate them gives the wrong view of how Bitcoin behaves in the real world. "Everything is connected—crypto doesn't sit outside the system."

That changes the conversation. Bitcoin stops being a fixed thing to define and becomes a player in a changing system. Its behavior depends on the situation, and it reacts to liquidity shifts, to swings in investor mood, and to policy moves that might not even mention crypto but still change the environment. That doesn't erase Bitcoin's features; it just puts them in context.

Outcomes come from both design and interaction. One feature Agne sees now is lasting uncertainty— not a momentary blip that returns to normal, but a continuous state fed by overlapping problems— trade tensions, geopolitical friction, changing rules, and big macro imbalances building up. They do not arrive sequentially; they accumulate as the system continuously adjusts.

In calm times, people optimize, make choices based on fairly predictable conditions, create models with some confidence, and place capital with long-term plans in mind. In unstable times that confidence breaks, decisions go defensive, people try not to lose instead of trying to win, and growth takes a backseat to protection.

That is where Bitcoin begins to look different. Historically, gold and property have been seen as hedges when things get rocky. These are not perfect hedges, but they are steadier than many other bets. Bitcoin is entering that conversation, not because everyone agrees with what it is, but because of how people use it. Faced with uncertainty in older markets, some investors shift into Bitcoin as part of managing exposure. As Agne notes, market uncertainty pushes people toward assets that can act like safe havens, and Bitcoin is being tested there. "Market uncertainty pushes people toward assets that can act like safe havens—and Bitcoin is being tested in that role."

What matters is not whether Bitcoin is a perfect safe-haven, but that people sometimes treat it as one. In complex systems, how people behave often comes before formal labels. If enough people use an asset in a certain way, that use shapes its role. Repeated choices create patterns, and those patterns change how the entire system looks.

That process is not straightforward because Bitcoin is not consistently a safe-haven nor purely a risk asset. It can act like both, depending on the context. In speculative runs, it moves with high-risk assets, driven by momentum. In uncertain periods, it can draw capital as people look for alternatives to traditional tools. That mix makes analysis harder, but it also shows a system still being defined by use.

To see how that plays out, you must follow capital flows. Money doesn't sit still; it chases incentives and avoids risks. "Capital moves where conditions make sense—it doesn't stay loyal to any one asset." Those flows respond to macro factors far beyond crypto. Interest rates, bond yields, currencies, and policy moves shape allocation. When those factors change, the effects reach traditional markets and crypto.

Take Japan, for instance. It's been central to the carry trade, where investors borrow at low rates and invest where yields are higher. That was a source of global liquidity. When Japan's conditions shift—higher bond yields or policy changes—the incentives that drove those trades can flip. Money that flows out can be pulled back, and markets tighten. That is not a crypto-only event, but it still directly affects crypto. If liquidity shrinks, risky assets feel the squeeze; and if capital gets reallocated, Bitcoin's role in portfolios can change.

The same logic applies to geopolitical moves. Trade disputes between big economies add uncertainty. Sudden policy decisions make outcomes difficult to model. Even smart players can struggle to guess the next step. The result isn't just price bumps; it's a wider instability that changes behavior across markets. In those conditions, the idea of a black swan matters not as a prediction, but as a reminder: not all risks can be foreseen. Systems become more complex, and then unexpected outcomes come. When they hit, they usually expose weaknesses that were hidden in normal times.

For Bitcoin, its role is constantly tested in real-world conditions where multiple forces interact. Does it keep value under stress? Does it remain usable? Does it attract or lose capital? The answers shape its place in the system.

Agne does not claim firm answers, and she does not try to. She looks at the conditions that produce answers by watching how Bitcoin reacts to pressure, how people use it, how money moves through it, and how participants act. She spots patterns that theory alone would miss. That changes how we study Bitcoin. Instead of pinning down a single definition, it's more useful to follow how its role shifts. "Instead of asking what Bitcoin is, it's more useful to watch how it behaves under different conditions."

Something mostly seen as speculative can be treated as a hedge in some situations. What institutions once ignored might become part of their allocations. These moves don't come from one event; they come from experience piling up. Over time, repeated exposure changes behavior. Investors who ignored Bitcoin may look at it, and institutions that stayed away may build ways to engage. Small steps add up and slowly reshape the system.

That ties Agne's view back to the book's bigger frame. If Sung shows how Bitcoin fits into current structures, Agne shows how those structures act under strain and where Bitcoin sits in that behavior. Structure gives access, while pressure shows function. The system looks stable until it's tested; then its real traits appear.

That testing is happening now, as the global environment is not returning to a stable baseline. It keeps evolving, pushed by hard-to-predict forces and in a world where assets and systems get judged by how they perform under stress. Bitcoin, seen this way, isn't defined by what it claims to be; it's defined by how it behaves when the world around it is strained. Bitcoin's role is forming not in isolation, but in response to factors that are still unfolding.

One straightforward way to see how systems behave under pressure is to watch how different groups react to the same conditions. In theory, markets act as one thing, but in practice, they are a bunch of actors with different incentives, limits, and time horizons. Those gaps grow when uncertainty rises because they shape how capital moves.

Retail traders often react to volatility by following sentiment and recent price movements. Narratives spread fast on media and social channels, and that changes behavior quickly. In times of uncertainty, that looks like sudden selling in a downturn or aggressive buying in a recovery. Both make the volatility of retail investment bigger because it is driven by judgment, limited information, shorter time horizons, and no formal risk rules.

Institutions behave differently, but they feel the same pressures. They have mandates and fiduciary duties, and they need to explain moves within formal structures. When things get uncertain, they don't just snap...they reassess, run allocation models again, look at correlations, and think about how assets behave under stress. Institutions are limited by slower, more deliberate processes and are not driven by whatever mood is trending.

Bitcoin sits where both retail and institutional behaviors meet. Retail flows can push prices around fast while institutions push a different logic; both combined can add a kind of stability or another way of valuing it. That dual pull makes Bitcoin's behavior complicated—not purely speculative and not fully institutional—landing somewhere in-between.

Different types of capital interact, sometimes boosting each other and sometimes pulling in opposite directions. This dual pull shows up most when markets are stressed, as liquidity tightens and institutions cut exposure to risk assets. That puts downward pressure across markets as retail sees falling prices and may sell, which speeds the drop. Other players look at the same moves and see buying opportunities and step in. So, you don't get one unified reaction; you get layers of responses based on different frameworks. That's why calling Bitcoin a "safe-haven" is messy.

A safe-haven keeps value or gains in stressful times. Gold has done that, and some government bonds too. Bitcoin doesn't do that reliably yet...sometimes it falls with other risk assets, and sometimes it decouples and draws capital even as other markets weaken.

Agne's view lets you be flexible about this. Bitcoin can act as a hedge in some situations, especially when people worry about currency stability or systemic risk. It can act like a risk asset when liquidity is loose and speculation is high. Those roles can co-exist, but which one dominates depends on the environment. Roles in complex systems are assigned through use. An asset becomes a hedge because people use it that way repeatedly under certain conditions.

That behavior then shapes perception, and perception changes future behavior. What starts as an experiment can become an expectation. Feedback loops speed this up because market moves shape narratives, which shape perceptions that shape allocations.

If Bitcoin holds up during a crisis, the story that it can hedge grows stronger, bringing more capital and supporting performance. If it falls sharply during times of heavy stress, the opposite happens and people reassess. Those loops aren't controlled and they aren't linear. They make cycles of reinforcement and doubt. That's why Agne avoids hard labels for Bitcoin and looks at how it behaves in different situations and how behavior changes its role in the system.

Time horizons matter. Short-term moves can fool you because they often respond to single events, while long-term patterns show adaptation. As Bitcoin goes through cycles of stress and recovery, its behavior during those moments becomes more telling. People build expectations from what they see and then position themselves differently next time. This slow pattern recognition shapes the system; it doesn't make things certain, but it reduces ambiguity.

People might not agree on what Bitcoin is, but they often agree on how it behaves in certain conditions. That shared sense shows up in allocation choices, risk models, and market strategies. The system keeps changing as new players join with different motives. Rules and regulations shift how assets are traded and held, as technology changes how people access Bitcoin. Each factor alters behavior and the system itself.

Any analysis has to adapt because fixed models get outdated fast. Agne focuses on conditions, not conclusions, as she watches how the system reacts to pressure and frequently updates her view. She doesn't force a static framework on a moving target. Pure ideology or pure technology doesn't hit the bullseye. Ideology says what a system should be, technology says how it works; and neither shows how it behaves inside a messy, connected world.

Behavior comes from interaction, which is shaped by outside forces you can't fully control. Bitcoin is made by its environment as much as by its design. It reflects who uses it and how those people act when things get uncertain. That doesn't make it any less unique, but it puts it in a wider context that matters for understanding its role.

As global pressure builds, this view becomes more useful. Systems under stress reveal structure and weak points to Bitcoin's ability to function in that environment—whether as a hedge, a risk asset, or something else—and question how it fits into bigger financial frameworks. That leads to key questions about Bitcoin: What can you actually use in the real world? Which features work outside theory? Which parts translate into reliable functions when markets, users, and integration test them?

Because once a system faces pressure, the gap between possibility and practicality is glaring. And that gap is where the next layer of understanding appears. Over repeated cycles, one big change is not in the asset but in how it fits into formal frameworks. Edge cases become things for which you must account, not because uncertainty is gone, but because something kept showing up across different conditions. Bitcoin is in that stage now.

Bitcoin is not only an experiment or a speculative play; it's being treated as a component in allocation strategies, even if allocations are small or conditional. That shift doesn't come from one event; rather, it comes from repeated exposure, tests alongside other assets, and recorded behavior that participants use when making decisions.

Institutions are deliberate about this. While retail traders can move fast with little justification, institutions need risk models, compliance checks, and governance to add a new asset. Bitcoin's path into these frameworks is slow and shaped by both internal review and outside validation.

Agne points out that the wider environment matters. When markets are stable and traditional assets do well, there's no reason to add complexity. When instability rises, correlations change, traditional hedges underperform, or policy adds uncertainty, institutions must rethink assumptions. Things once seen as peripheral look relevant. And Bitcoin's relevance is conditional, depending on relative behavior, liquidity, and fit within risk frameworks. But once it's in the analysis, it's hard to ignore. Even a small allocation needs monitoring and modeling, and that attention normalizes it. Normalization doesn't mean full acceptance; it means being part of the conversation. Bitcoin doesn't need everyone to agree to matter to institutions. It just needs to be relevant enough to consider. After that, its behavior is tracked, its role debated, and its presence affects decisions, even indirectly. That brings a different stability: not price stability, but stability in attention.

Bitcoin stops vanishing with market cycles and stays on the map. That persistence helps it become more embedded, and the more it's watched, the more structured participation becomes. Integration doesn't erase volatility or uncertainty, as Bitcoin still reacts to liquidity shifts, investor mood, and macro forces. What changes is how people interpret those moves. Instead of dismissing volatility, they study it and build it into models. That's a small but meaningful shift.

An asset written off for volatility is unreliable; and one whose volatility is studied becomes complex. Complexity doesn't stop participation; it makes it more structured. Agne's lens meets the institutional side here. By looking at conditions, not conclusions, she shows how assets move from the edge toward the core. That movement is not driven solely by narrative or technology; it is shaped by repeated interactions under different conditions, where people update their views based on what they see. Over time, that creates bottom-up, uncoordinated, distributed collective learning from players reacting to the same environment in different ways.

Some act fast and experiment with Bitcoin early, while others wait for more cycles and more evidence. Both paths shape the overall direction. This distributed learning allows the system to adapt without a single point of agreement. It also keeps Bitcoin's evolution tied to real-world conditions, not abstract labels. If people keep testing, allocating and adjusting around it, Bitcoin will keep evolving within the system. But evolution is not only additive; it also introduces limits.

As Bitcoin gets folded into institutional frameworks, it's exposed to the forces that shape those frameworks. Rules, reporting, and risk management change how it's used. They don't alter the protocol, but they change access and which forms of participation win out. And the tension between structure and environment gets sharper. Structure enables participation within set systems, while the environment shapes how that participation behaves under pressure. Together they decide Bitcoin's practical role.
Agne lines that up by focusing on the conditions that drive behavior. She doesn't treat Bitcoin as

isolated or reduce it to one use; in contrast, she watches how it sits inside a shifting network of relationships. That position decides its relevance, its use, and what it means in the bigger system. As global conditions shift, that position will keep changing. As new instabilities pop up, old ones grow or shrink, and participants shift, Bitcoin's role will keep reflecting both its internal traits and the external forces around it.

We now return to a transition point. Studying how a system behaves under stress gives insight into roles and changing importance and shows how people respond to uncertainty, how capital moves, and how assets get rethought. But it leaves a practical question you can't ignore as systems mature.

What actually works in practice—not by theory or narrative, but what holds up under real-world constraints?
Which use cases survive testing by users, markets, and integration?

Those questions shift the focus from watching to doing. And when systems face pressure, existence is not enough.

They must function.

Execution and Reality

There comes a point in any system's life when possibility stops being the primary concern and reality takes over.

Early on, we judge systems by what they might become, putting up with limits and forgiving a great deal because we believe in the promise. But that patience runs out as a system grows, more people join, real money gets involved, and real consequences begin to appear. The rules change, and people stop caring about promises and start caring about what works in uncontrolled situations. "At some point, you stop asking what could work and start asking what actually does."

This is the world Mark Højgaard works in—building and running systems that must hold up in the real world. He studied economics and philosophy and then spent decades as an entrepreneur. That mix makes him question assumptions while staying focused on what produces results. It allows him to engage with crypto ideas while still testing them against actual products and services.

In 2014, he co-founded Coinify, which sits where crypto meets legacy finance. It is neither an exchange nor a wallet. It acts as a bridge that enables payments and provides on and off ramps so people can move between crypto and fiat. That positioning was deliberate because scaling crypto is not just a tech problem; it is an operational problem, too.

Systems must connect, and value must move across boundaries to fit existing financial rules and expectations. From that viewpoint, the gap between story and reality becomes obvious. Bitcoin was pitched as peer-to-peer money—fast, cheap, and without intermediaries. In practice, things look different.

Mark sees that Bitcoin still leads in transaction volume in some contexts, but it does not behave like a payment tool when you measure it by real-world commerce. Fees, price swings, and settlement quirks add friction—enough friction to make everyday transactions awkward. Bitcoin can show high transaction volume and still not be the best tool for payments. That difference matters. "High usage doesn't mean suitability—especially in payments."

There is a gap between what a system can do and what it fits. Capability alone does not create adoption. Suitability does. How well something matches a use case drives steady use, especially payments, which come with strict constraints. Transactions need to be quick and costs must be predictable. The unit of account should not jump while the payment is happening.

These are not abstract. They are basic expectations for any system that wants to handle commerce.

Mark thinks Bitcoin often misses those marks. That is not a failure of Bitcoin; it is a mismatch between its traits and the needs of payments. Seeing that mismatch is not rejecting Bitcoin; rather, it is redefining its role.

Instead of forcing it into day-to-day payments, Mark treats Bitcoin differently. It becomes more of a store of value, a scarce asset with specific properties, rather than the primary medium for everyday purchases. That opens space for other tools, such as stablecoins. Because stablecoins keep value tied to fiat, they remove a major source of friction in payments. With no wild price swings, stablecoins are easier to use for commerce.

The uptick in stablecoin use, in Mark's view, is not hype and simply happens because they work where stability matters. "In payments, stability isn't optional—it's the baseline." That practical view recurs throughout his thinking.

Early crypto favors novelty, new ideas, experiments, and pushing limits. Over time, the focus tightens as systems are judged by whether they deliver steady results. Features that do not translate into reliable function get dropped. Use cases that cannot scale are reworked or abandoned. You might not see the work, but it happens all the time.

Mark lives inside that filter. His job is choosing which possibilities can become systems that work under real constraints and deciding what to back and what to leave out. "Not everything that works in theory survives real constraints." Meme coins get attention and trading, but they do not meet the standards for payments with big partners. Dealing with them brings risks that Coinify cannot accept, so the meme coins are treated with caution or not at all.

Selectivity is not a matter of taste. It is necessary. When you work with major payment providers, global brands, and regulated frameworks, the margin for error is tiny. You must consider reputational harm, compliance, and the effects on partners who count on the system. That forces discipline that is often missing in more speculative corners. Discipline filters things, cuts what cannot survive scrutiny, favors the steady over the new, and favors reliability over wild experiments.

Over time, that filtering shapes the system, not by killing innovation, but by steering it toward what can actually be used. This is where Mark's view meets themes from Sung and Agne. Sung shows how structure creates access. Agne shows how systems behave under pressure. Mark shows what's left when you apply both: what actually works when you structure something and then test it, and what survives when theory hits the real world. The answer is not everything. It is a subset of a smaller, tighter set of use cases, tools, and behaviors that hold up across different conditions.

That narrowing is not weakness; it is maturity. As systems evolve, they become more defined, and more about what they exclude than what they include. That iterative process continually gets tested by market forces, rules from regulators, and participants testing limits.

Mark's role is where ideas meet execution, concepts are turned into systems, those systems are judged by performance, and performance determines what stays. If the theory versus reality gap is most visible anywhere, it is in the user experience. A system can be clever, elegant, and even make economic sense, but if people cannot use it easily and reliably, then adoption stalls not because the design failed in principle, but because real people interact with technology in messy ways.

Mark's approach starts from that. The job is not to convince people that crypto is valuable, but to make crypto usable without asking people to change how they behave. Adoption rarely happens if users must adapt to the system; it happens when the system adapts to them. Payments make that especially hard. Someone paying online or in a store wants things to be instant, obvious, and reliable, with little tolerance for complexity and even less for uncertainty. A failed or slow transaction destroys trust.

Traditional payments set a top bar over decades. Credit cards, digital wallets, and payment gateways created expectations that new systems must meet. Raw crypto often falls short of that bar. This is an observation, not a dig. Dealing directly with blockchains requires attention and knowledge that most people do not have. Private keys, wallet screens, fees, and network delays add friction that was acceptable to early adopters but is a real barrier for everyone else.

Mark pushes infrastructure, not by simplifying the blockchain itself, but by adding layers that hide the messy parts—payment gateways, on and off ramps, compliance checks, and middleware that sit between the user and the protocol. That infrastructure turns complexity into something that feels familiar and predictable.

That translation is difficult because you must align tech integration, rules compliance, risk controls, and the user interface. Each piece comes with limits. The trick is to join them without breaking the experience. When it works, users can deal with crypto without knowing much about it. Call it invisible infrastructure. Users do not see the gears; they see the result: a payment goes through as the transaction completes and value moves. From the user's perspective, it works. From the operator's side, there is a lot of complexity being managed. That dynamic makes middleware crucial.

People mostly ignore middleware when talking about crypto. Conversations focus on protocols, tokens, and apps. But middleware shapes a lot of how the system works as it links different parts, manages how they talk to each other, and keeps things running across boundaries. At Coinify, this is explicit. The company sits between crypto and fiat, moving value back and forth. That means technical integration, following rules, and meeting the standards of big payment service providers.

These are large global firms with strict requirements. You must meet them without exception, which is the price of playing. This is where trust matters; not some abstract trust, but practical trust in how things run. Partners need to believe transactions are handled right; regulators need to believe compliance is followed; and users need to believe their money is safe and the system won't surprise them. Trust requires more than technology. It takes processes, controls, and accountability.

That sounds like old finance concepts, and they matter for scale because without them, the system stays small. Only people willing to take more risks and handle complexity stick around. At that point, adoption becomes the central question.

To reach beyond early adopters, crypto must deal with users and institutions that expect different things. These people care about usefulness, reliability, and risk, and not ideology. If the system wants them, it must meet those needs. That pulls compliance into the story.

People often treat compliance as a limitation because it adds steps, cost, and friction, but it also builds a framework. Compliance lets participants act with some confidence they wouldn't have otherwise. Mark sees it that way, not as a hurdle to dodge but as a condition to satisfy. It's part of the environment you operate in.

If you want to plug into mainstream finance, you must engage with compliance, especially when big partners have their own legal duties and can't take on extra risk. Compliance shapes which assets and use cases get supported because not all tokens are equal. Higher-risk assets—volatile tokens, opaque ones, or meme coins tied to speculation—are less likely to be integrated where stability and trust matter. That narrows the choices, but it makes the system more reliable.

Reliability becomes a competitive edge as it separates tools you can use in real life from ones that remain experimental. Over time, that matters. Reliable systems attract more users, more partners, and more capital, while less reliable ones get left behind. This happens quietly but constantly.

Companies build infrastructure, institutions move capital, and users pick platforms. Each choice pushes the market. Paths get reinforced while others fade. So "what actually works" turns into more than a practical point; it becomes a force.

What works gets used; what gets used gets built on; and what gets built on shapes the system. That loop explains a lot of what Mark talks about. It shows why some stories die out and others rise. It explains why promising tech cannot scale and why systems that look less flashy can win if they match real-world needs. Other factors complicate this—platform collapses, hype cycles, and waves of innovation all change how people behave.

Mark points to meme coins and speculation as problems. They bring short-term activity, but they also add risk that erodes trust and slows long-term adoption. Fixing that needs two things: market discipline and regulation. Market lessons come from experience. People who lose money betting on hype often become more cautious. Over time, that can make choices smarter and reduce herd behavior.

Regulation tries to set boundaries and rules to spot fraud, raise transparency, and protect people from big systemic risks. Regulation won't wipe out danger, especially in a fast-changing space, but it can cut how often and how badly things go wrong. Mark thinks you need both. Markets alone don't always force discipline when speculation pays off. Rules alone can choke new ideas if done badly. Finding the balance is hard and ongoing.

Crypto's life cycle looks familiar. Early on, it's all exploration as new ideas get tried and tested. Later, things consolidate as winners scale and get folded into bigger systems. That shift is bumpy and filled with tension. Some people push for newness and flexibility, while others want stability and control. Both sides matter and lead to different results.

Mark tends to the stability side, not because he hates innovation, but because systems need steadiness to work at scale. That doesn't fix the tension; it highlights it. Evolution, not a straight path toward one thing, is shaped by clashing priorities. As this moves on, a key split appears: what's possible versus what's practical. Possibility fuels exploration, but practicality drives who actually adopts something.

The winners bridge both. They're recent enough to offer something useful, but solid enough to fit into real-world limits. That's Mark's space. It's where the next phase gets decided. Once a system proves it can run actual use cases, plug into existing infrastructure, and remain reliable, the question shifts. It stops being about whether it works and starts being about what it stands for and what it should become.

Here selection kicks in. Early stages add options fast as people try things and push boundaries. But since you can't keep adding forever, the field narrows. Selection doesn't happen by decree; it happens by usage. Systems that get used, integrated, and relied on keep evolving. Others fall away. Not because they're wrong, but because they don't fit the limits of real environments.

You don't always notice while it's happening, but over time the pattern is clear. What works, stays; what doesn't, fades. Mark sits in the middle of that selection. His job isn't to create more possibilities, but to filter them. He takes the noisy set of crypto ideas and checks them against a different rule set—usability, reliability, integration, and not ideology. Can it work where failure has actual costs?

That filtering is active with ongoing checks. Every integration decision, every asset supported, and every product choice is a call about what can be trusted in a structured setting. Those calls are shaped by partners, regulators, and users who won't tolerate mistakes. Being wrong is painful and causes lost trust, broken relationships, and lost money.

That's why Mark's view matters. He doesn't just watch what works; he commits to it. "In the end, what gets used is what defines the system." When a system goes into a payment network or gets offered as a reliable way to transact, it moves from idea to commitment. The room for error shrinks fast. Commitment brings discipline as it makes outcomes more important than stories. In a world of hype and rapid change, that discipline balances things out. It doesn't stop the experiments. It just means only some experiments end up scaled and used in the real world.

Over time, this creates a split between the flashy, headline-grabbing layer and the operational layer. That gap is a defining trait of a maturing crypto space. On the surface, everything looks broad and new, but underneath, things get tested against real requirements, and many don't pass. They don't vanish entirely, but their influence drops compared to the ones that hold up.

You see this most when things get tense. In calm times, more ideas survive, and people take risks. When pressure rises, tolerance falls. Systems that can't keep working under stress get exposed, and users move to those that show resilience.

This is where Mark's view meets Agne's. Agne shows how systems act under pressure. Mark shows which parts still work when pressure hits. Together they point to a two-step process: exposure, then selection. Pressure reveals flaws, and selection decides what remains. This keeps going as fresh waves of users, new rules, and new technology add variables to judge. The system gets reshaped not by a single route but by many choices that push it toward certain paths, with the subtle result being standardization.

When some systems prove reliable, they set expectations, and new projects follow proven models. That can speed up adoption because people like familiar things, but it can also shrink experimentation. Stray too far from the norm and you take extra risks. That tension exists in every maturing field. The trick is to stay flexible enough for new ideas while keeping core functions steady.

Mark leans toward steadiness, not as the result, but as a step needed for scale. Without it, you don't integrate, and without integration the space stays fragmented. Fragmentation is fine early on, but later it blocks growth. Users don't want to juggle many incompatible systems. Partners don't want to wire into platforms that may disappear. And institutions don't put money into places that aren't consistent.

Fixing fragmentation needs coordination even in decentralized setups. Standards, common practices, and the pull of entities that sit between parts of the system all help.

That brings us back to middleware that ties systems together and enforces shared processes. It does not erase differences, but it creates a compatibility layer that allows systems to inter-operate. That's key to scaling beyond small niches.

Mark's role in that layer gives him a clear view of consolidation. He sees what's built and what actually gets integrated. He sees not just possibilities but what's used. That view defines his lens, as he isn't focused on the far edges where people keep experimenting. Instead, he's watching the center where decisions stick.

As the space matures, the center grows louder. It shapes how new people join and sets the bar new systems must meet. It decides which use cases catch on and which stay on the side. That concentration of influence isn't classic centralization; it is convergence. People move toward systems that work, and those systems grow stronger.

This adds another tension. Sung shows how structure opens access. Agne shows behavior under stress. Mark shows how these push the system into something that can work at scale. But narrowing has costs because it decides what's inside and what's outside. And at that boundary, new questions appear, not about whether things function, but about intention.

What should the system keep? What should it push away? What must not be lost while making it work?

Those questions don't come from doing things alone. They emerge from principles and beliefs about what the system should mean.

And in that meeting of practice and principle, the next lens becomes necessary.

Preservation and Drift

At some point, a system stops being about how it works—or whether it works—and becomes about what it is becoming. That shift doesn't just come from technical limits or market pressure. It comes from something quieter—the pile of decisions that, over time, shape a system's character. "Bitcoin isn't just code—it's a set of principles, and those principles can be lost over time."

By then, most early doubts are gone, the mechanics are understood, use cases are clearer, paths for participation are established, structure exists, and pressure has already been applied. What remains has survived. It works, but working is not the same as staying true.

A system can keep working and still drift away from what mattered at the start. It can grow, get integrated, be widely used, and lose the traits that made it distinct. That is not a failure, exactly; it is a consequence of success.

This is where Vlad Costea shows up. He doesn't begin with what Bitcoin can do. He starts with what Bitcoin is supposed to be. His relationship with Bitcoin was slow. He did not treat Bitcoin as a religion or elevate its figures into heroes, as is common among anonymous maximalist accounts on social media channels. At first, he hesitated, avoided it even. Like many, he heard the loud stories—illicit markets, fringe activity, and risks that felt bigger than curiosity.

That hesitation matters because it's a psychological barrier, not a technical one. Early Bitcoin didn't invite casual study; it felt like something you approached carefully, or not at all. Time changed that, and so did investigation. When Vlad returned years later, he did so with intent—he wanted to test the claims about decentralization, no central control, and a network with no single authority. You can't take those on faith, so he looked for control, for the person behind it, for the roadmap steering it, and for the structure that governs it.

What Vlad found surprised him. There was no organization, no authority, and no formal governance. Bitcoin isn't unstructured—there are developers, maintainers, and contributors who influence direction—but no single entity defines it, no central authority calls the shots, and no roadmap guarantees where it goes. "There is no one in charge—and that's the point."

What exists is a process that is open and messy, requiring contribution and argument, and driven by slow consensus that evolves by iteration and by friction more than neat alignment. Vlad calls it "anarchy on purpose", distinguishing it from chaos, which is the absence of imposed order.

Coordination is present because people participate, not because someone controls it. Decisions stick because others accept them.

That is the cypherpunk core. It doesn't chase efficiency or scale first (ease of use comes later); it values something else far more important: Sovereignty.

Sovereignty is the ability to join without asking permission, to hold value without intermediaries, and to move value without outside control. These are not features; they are the foundation. "If you need permission, it's not really yours." And that foundation gets exposed as the system grows. The forces that make scale possible — structure, integration, standards — don't line up with the forces that protect sovereignty. These forces introduce intermediaries to simplify use, increase control, hide complexity, and shift responsibility. That is not an accident. It's what happens when something moves from the edge toward the center.

Sung shows how structure opens access. Agne shows how pressure shapes behavior. Mark shows how reality filters outcomes. Vlad shows what's at risk when all that happens.

The risk isn't that Bitcoin stops working; the risk is that it keeps working, but in a way that differs from what was originally intended. You see that in how people view Bitcoin. For some, it's mainly a financial tool, a scarce asset with a fixed supply meant to store value. Their worry is security. Does the network run? Is the supply fixed? Can it be trusted long term? That's the Austrian view.

For others, it's more fundamental—a tool for autonomy and a way to operate outside normal finance. Their worry is use. Do users control their keys? Are transactions private? Is participation really permissionless? That's the cypherpunk view.

The Austrian and cypherpunk views overlap. But they're not the same. As Bitcoin changes, balancing these two views gets harder. Vlad's worry is visible and concrete: as Bitcoin ties into mainstream finance, the story shifts.

Talk leans toward price, institutions, and fitting into rules. That brings money, attention, and a kind of stability that wasn't there before, which is not all bad, but it changes incentives. If you hold Bitcoin on custodial platforms, you don't control your keys. If you route transactions through intermediaries, the system feels less direct. And if you focus on price, other parts get less care. Not a collapse; a transformation. "The danger isn't that Bitcoin fails—it's that it succeeds in the wrong way."

That transformation raises a simple question:
When does the ease of access erode autonomy?

There is no simple answer, as the trade-offs are real. Making Bitcoin easier usually means hiding details; hiding details creates intermediaries who bring trust, which brings back the dynamics Bitcoin was meant to reduce. Vlad doesn't say systems shouldn't change. They must adapt to survive, but adaptation needs limits— not limits set by a central power, but limits set by principle.

Vlad's take on Bitcoin maximalism fits here. This view, which for many resembles a religion, is not a rigid creed, but more like resistance. It is a way to keep sight of what Bitcoin is supposed to protect even while it changes. Maximalism is not only about rejecting alternatives; it is about keeping the system's edges intact; making sure experiments don't water down the core; making sure comfort doesn't beat sovereignty; and making sure integration doesn't become dependence. It can sound extreme, sometimes confrontational, and dismissive of other approaches, but under that stance is a constant worry.

As Bitcoin becomes more acceptable, it can also lose what made it distinct. This worry goes beyond Bitcoin. It applies to the entire ecosystem. Altcoins and experimental platforms aren't automatically bad because they let people try things Bitcoin can't yet. They widen the possibility, but they add noise by creating stories that fight for attention. They set incentives for short-term gain and form structures that might not last.

Vlad says you need to sort that out. Not every innovation is progress, and not every new user base matters the same. Some growth is a distraction, and over time, distraction weakens focus. That's why Bitcoin remains central to him. Not because it's perfect, but because it's persistent, doesn't change easily, and resists quick fixes. That resistance acts as protection, preventing the system from being reshaped too quickly. Changes are tested and argued repeatedly and stick only when they survive that process.

That friction slows things, but it filters them too. The toughest ideas survive, and the ones that really matter get in. That matches, oddly, what Mark calls selection. But Mark's selection is practical, while Vlad's is philosophical.

What should be allowed?
What should be pushed back?
What must stay the same?

Those aren't technical questions; they're ideological, and they matter. Without that debate, a system becomes purely functional, efficient, scalable, integrated, and useful to existing structures, but no longer distinct. That's the hard tension. A system that fits perfectly into institutions might stop challenging them. A fully integrated system may stop offering an alternative. If ease wins everything, people stop needing to understand. And without understanding, the original purpose fades.

Vlad doesn't fix the tension, but he keeps it alive. He insists that as Bitcoin changes, there must still be a link back to what it was meant to do, not only in ideas but in practice. That practice includes letting people act independently, keeping a system that is hard to control, and making sure the core stays reachable under all the layers. That's not guaranteed; it depends on users who engage directly, developers who put those principles first, and community members who remember why Bitcoin exists.

As Bitcoin moves on, that memory matters more. Systems don't lose purpose suddenly. They drift slowly through small choices and tiny shifts that pile up and steer things elsewhere.

Spotting drift is the first step. That's what Vlad's view serves as a reminder of what must not be lost. "If you don't hold your keys, you don't hold your Bitcoin." What makes this tough is that the change doesn't start as a fight, but arrives as alignment with the same things that make Bitcoin easier to use—custodial services, exchange-traded products, ties to banks—which also make it more visible, more liquid, and more widely accepted.

For many, that looks like success. It convinces people who would otherwise ignore it, brings capital, and creates a larger user base. But fitting into existing systems is not neutral because it changes incentives. Hold Bitcoin through institutional products, and the interaction is mainly financial. In that case, you don't run a node, verify transactions, or manage keys; you interact with a representation of the asset. Prices might move the same, but control doesn't.

That difference seems small, but it is significant in system terms because over time the gap grows. If most people use custodial or institutional layers, fewer deal directly with the protocol. Knowledge about securing keys, verifying transactions, and running infrastructure concentrates in fewer hands. That doesn't break the network immediately, but it alters who has responsibility.

Responsibility moves from the individual to the intermediary. While that shift is efficient, cuts friction, and lowers barriers, it also recreates the dependence Bitcoin set out to reduce. That's the structural tension Vlad answers. Not that institutions are evil; just that they change the balance between sovereign control and ease.

The Austrian view accepts trade more easily. If Bitcoin's job is sound money, a scarce verifiable asset, then whether you hold it directly or through a bank matters less if the network is secure and supply stays fixed. Banking looks like a natural step as money scales. The cypherpunk view disagrees. For them, the way you interact matters. If you don't control your keys, you don't control your money. If intermediaries can watch or block transactions, the promise of autonomy weakens, even if the protocol stays the same. Not an argument against scale; it is about how that scale is achieved.

The two views also aim at different things. The Austrian view aims at monetary integrity, while the cypherpunk view aims at operational sovereignty. Both have value, but pull the system in different directions. Vlad's maximalist view sits within that tension, without a single fixed definition, but a contested space where meanings compete.

Some see maximalism as an economic bet—Bitcoin will beat everything else, so you should hold only it. That view, otherwise known in the industry as toxic maximalism, reduces things to price. Others see Bitcoin as technical—real innovation should end up on the Bitcoin blockchain. That view allows experiments but wants the ultimate home to be Bitcoin.

A third version matches Vlad better: maximalism as preservation. Resist dilution; ask if new features or integrations fit the original aim; and choose restraint over constant expansion. People often misread this as rigid or hostile, but it often acts as a counterbalance. It slows changes that might come too fast and forces each change to be judged for utility and for how it lines up with core properties. That resistance can be uncomfortable because it creates friction in the community, causes debates that feel unproductive, and sparks arguments that seem adversarial. But without that friction, the system would move faster. And faster is not always good.

When the cost of error is high, you need caution as well as speed. Bitcoin is built around that idea: changes are hard, consensus is hard, and upgrades take time. People call that a limitation, but Vlad calls it a safeguard. Because there is no central authority, once decisions stick, they're very hard to reverse. Mistakes don't get fixed quickly, so it is better to avoid them. Therefore, most experimentation happens outside of Bitcoin, in alternative systems and in places where failure is less costly.

Vlad isn't blind to the value of those places. He says they matter as they let you try things that wouldn't fly inside Bitcoin. But he also sees that those environments have different incentives. Many alternative systems favor early participants. They use pre-mines, allocation schemes, and governance that concentrate influence as they lean on marketing and hype to pull users in. That is not always bad, but it changes the game. It pushes more toward venture capital and speculation than toward the principles behind Bitcoin.

That's when the language gets sharp. Words like "shitcoins" show up. That is not a neutral stance, but a way to reject everything that is not Bitcoin. From the outside, it sounds harsh and even unhelpful, but from Vlad's point of view, it has a job—it draws a line, not a technical line, but a conceptual one. It separates things aligned with Bitcoin from things that aren't and reminds people that not every innovation is the same. Some ideas are neat technically, but not things Bitcoin should adopt.

Not everyone accepts that boundary...and they shouldn't. Alternatives create competition, which often drives improvement. But Vlad thinks Bitcoin shouldn't try to compete on every front. Its role is to keep a specific set of properties that others don't prioritize.

That's where time matters. Bitcoin doesn't move at the speed of other projects. It doesn't add features fast, and it doesn't chase market trends. That slow pace can be frustrating, especially when the rest of tech never stops. But it also creates stability—not price stability—predictability. The system won't change unexpectedly.

The Bitcoin core stays the same for long stretches. That predictability allows people to build infrastructure on it, move capital, and connect it to other systems. Without that base, none of it works.

Stability alone is not sufficient. A system that never changes can become obsolete, so there's a tension between keeping things and changing them.

Vlad isn't against change, but says it should be deliberate. Changes need more than short-term wins; they require long-run effects. This is where Vlad and Mark differ. Mark looks at what works, what you can use reliably in the real world, while Vlad looks at what should be kept, and what must stay intact as things evolve.

The two have different filters. Mark filters for functionality, while Vlad filters for principles. Together they mark the space where the system can move. If you push too hard for functionality, you lose what makes Bitcoin unique; if you push too hard for preservation, it can stall. That balance shifts as new people arrive, new use cases appear, and the world outside changes.

Vlad's stance isn't a conclusive answer, but it is something to revisit. Test it against new events without giving up the major concerns. The actual risk isn't that Bitcoin fails; it is whether it succeeds in a way that hides what made it valuable. Widely used, integrated, accepted... but no longer understood, no longer engaged with directly, and no longer used the way it was meant to be used.

This change does not happen overnight. It is a slow drift that, while hard to spot as it happens, becomes clearer later. The system still works, but it feels different. Paths that used to be open close off for most people, and the know-how to engage directly gets buried under layers.

Vlad's point is to notice the drift before it's too late, not to stop change, but to make change visible, keep the system anchored, and keep the original properties intact:

Accessible.

Usable.

Real.

The tricky part is that the drift rarely looks like a big
shift. It's a stack of small choices that each make
sense. Each new access point, each integration, each
abstraction nudges the system. None alone will
redefine Bitcoin, but together they add up.

Over time, you get a version that looks familiar,
usable, accepted, but is far from how it started. That
outcome isn't failure; it's success meeting reality. A
system that lives entirely outside existing structures
doesn't scale, won't get broad users, and won't fit
into the financial and tech fabric of the world. To
scale, it must interface with what's already there,
and those interfaces bring compromise. The
question is not whether compromise happens, but
how much can happen before the system loses its
core traits.

Vlad doesn't put a number on that limit. He offers a steady point of reference—a way to measure change against the original principles that often get forgotten once the system is running well. That's his role in the book. Sung turns Bitcoin into structured environments. Agne watches how it behaves under stress. Mark finds what survives in practice. Vlad keeps attention on what those processes might hide. He doesn't deny their need. He points out their cost, which doesn't show up in price or volume. It shows up in how people relate to the system.

When interaction goes indirect, when responsibility moves to middlemen, when understanding is swapped for convenience, control changes. Users may still get benefits, but they stop engaging with the properties directly. That shifts how the system is used and how it's understood.

Understanding isn't an abstract ideal. It's what lets the system survive across generations. Early adopters carry expertise about how and why it matters. As the system grows, that wisdom spreads thin. New users come in through simpler routes and meet a system already explained for them. That helps growth, but it builds dependency on the layers that explain things. If those layers dominate, they end up defining the system in practice.

That's when preservation matters. Not because the protocol will change immediately, but because access paths and experiences are shifting. Vlad pushes self-custody, running nodes, and direct participation not as a hobby, but as a way to keep a living connection to the base layer. He wishes to keep independent engagement possible as easier options appear, which ties back to time, not in terms of market cycles, but the long arc of system evolution.

Over decades, convenience, integration, and abstraction pile up. Systems get easier, faster, and more adopted, but they also make direct interaction rarer and the original traits more mediated. The point isn't whether Bitcoin will survive this (it probably will), but the question is whether its defining features remain reachable.

That's what resistance does: not to block change, but to slow it down. To make each shift visible, make tradeoffs clear, and keep the option of returning to core properties. Resistance isn't opposition; it's continuity that balances growth, so identity isn't erased.

That balance between growth and preservation doesn't come from one view alone but from all four. Sung makes it accessible. Agne shows how it behaves under pressure. Mark finds what works in the real world. And Vlad keeps what works from wiping out what matters. Together, they don't give one fixed definition of Bitcoin. They show its shape, which is a system molded by competing forces that don't cancel each other but sit in tension. That tension isn't a bug because it keeps things alive.

Inside that tension, the ultimate question forms: not what Bitcoin is, but what it becomes.

Bitcoin is Revealed

There is a natural urge, when faced with something as debated and complex as Bitcoin, to arrive at a single definition. After enough reading, analysis, and conversation, it feels as though it should resolve into something coherent—one stable idea that explains everything else. Most systems are taught that way: presented as complete, with fixed properties and clear roles.

Bitcoin resists that kind of neatness—not because it lacks structure, nor because it cannot be understood, but because it operates across layers that do not collapse into a single frame. Every attempt to define it captures something real. The result is not confusion, but fragmentation, as different participants focus on different aspects of the system.

Over time, those focal points harden into distinct interpretations. They do not cancel each other out;

they accumulate. Therefore, answers to the question of what Bitcoin is often feel incomplete, even when they are correct. It is money, infrastructure, a store of value, a network, and a system of rules enforced by code. Each of these is true. None is sufficient on its own.

It becomes more useful, then, to see Bitcoin not as a fixed object, but as a field of interaction. It was not simply designed and deployed. It is engaged with, interpreted, extended, and challenged. Each interaction reveals a different aspect of the system, and those aspects shape how it is experienced.

The central idea running through this book is simple: Bitcoin is not defined. It is revealed.

Revelation is not a single moment, but a continuous process shaped by use, pressure, resistance, and the environments in which people operate. It emerges through the incentives, constraints, and decisions that participants bring with them. What appears is not a single Bitcoin, but overlapping versions, each anchored in a different layer of the system.

The perspectives explored in this book do not aim to be exhaustive. Each isolates a force that shapes Bitcoin distinctly. Taken together, they provide structure without collapsing into a single answer.

Through Sung, Bitcoin becomes accessible. A system designed to operate outside structured environments is translated into them. The core is not altered, but the edges are interpreted. Infrastructure is built, interfaces are introduced, and compliance layers are added. These do not rewrite Bitcoin's rules, but they reshape how those rules are encountered. The result is a version of Bitcoin that fits within existing systems—accessible, scalable, but filtered.

Through Agne, Bitcoin is placed under pressure. It is within a broader environment of global markets, political decisions, and capital flows. In that environment, its role is not fixed. It shifts with liquidity, uncertainty, and external conditions, often behaving in ways that do not align neatly with its original narrative. The result is a responsive Bitcoin, shaped not only by design but by behavior under stress.

Through Mark, ideas are forced into execution. Use cases are tested against real-world constraints—technical limits, regulations, and human behavior. Systems are judged by reliability, usability, and integration. What cannot meet those standards fades, regardless of how interesting it may have appeared in theory. What remains is not always the most innovative, but what can function consistently. The result is a selective Bitcoin, defined by what it can sustain.

Through Vlad, the focus shifts to preservation. Bitcoin is traced back to its founding principles, and the question becomes what happens to those principles as the system evolves. As access expands, layers accumulate, and abstraction increases, participation changes. The system may continue to function, but the relationship between users and its underlying properties can weaken. The result is a contested Bitcoin—not in how it operates, but in what it represents.

These perspectives do not compete; they interact. Structure enables access, but also filters it. Pressure reveals behavior, but introduces volatility. Execution determines what survives, but narrows what is possible. Preservation maintains identity, but resists change. Together, they form a system of forces that is not unified but interconnected.

Understanding Bitcoin, then, is not about selecting one perspective. It is about seeing how these forces operate simultaneously and how their interaction shapes the system over time. This requires a shift— from asking what Bitcoin is, to asking how it is being shaped.

That shift is not abstract. It has direct implications for how Bitcoin is experienced.

Most participants do not interact with the protocol directly. They encounter Bitcoin through platforms and products that abstract away complexity. Their experience is defined by those interfaces—by how value is transferred, how custody is managed, and what features are available. That experience is not incorrect, but it is partial.

Under conditions of stress, behavior changes. Participants react to market instability, shift capital, and adjust positions based on how Bitcoin performs relative to other assets. These reactions create patterns, and those patterns influence how Bitcoin is perceived and used in subsequent cycles. In this way, behavior feeds back into the system.

Execution imposes further constraints. Not every idea translates into something that can operate at scale. Technical limitations, regulatory frameworks, and user expectations narrow the field. The system becomes defined by what works repeatedly. Innovation continues, but it is channeled.

Expansion introduces distance. As access becomes easier and participation grows, fewer users engage directly with the underlying mechanics. Convenience increases, but direct control can diminish. If the connection to core principles is not actively maintained, the system can drift—not through failure, but through gradual transformation.

These dynamics are not theoretical. They are already visible.

Bitcoin today is more integrated, more widely understood, and more accessible than it was in its early stages. It is also more layered, more complex, and more dependent on surrounding systems. This evolution is ongoing, and it unfolds through the accumulation of decisions—many of which occur far from the protocol itself.

Those decisions—how to build, regulate, allocate, and participate—shape direction over time. Not through coordination, but through accumulation. Patterns emerge as certain approaches are reinforced and others fade. Systems change in this way, not through consensus, but through repeated action.

The implications are straightforward. Bitcoin's future is not determined by design alone. It is shaped by use—by participants operating under different incentives, constraints, and perspectives.

This raises a set of questions that cannot be resolved immediately:

Will Bitcoin remain a system centered on direct, sovereign participation?

Will it primarily function as a financial asset integrated into existing structures?

Can both persist, or will one dominate?

These questions do not have fixed answers. They will be resolved gradually, through the same processes that have shaped Bitcoin so far—interaction, pressure, selection, and resistance. There will be no single moment of clarity, but a continuous unfolding.

Each participant contributes to that process, whether intentionally or not. Bitcoin cannot be defined once and finalized. It must be observed as it evolves. The shift is from viewing it as an object to understanding it as a process—one shaped by systems, by participants, and by the decisions that guide both.

When Bitcoin is described as being revealed, it is not because it is hidden, but because its nature is not singular. It emerges through interaction. Its meaning is not fixed; it is expressed differently depending on the context.

Understanding it requires continuity. It requires observing how it behaves across conditions and geographies. In the short term, Bitcoin appears volatile and reactive, influenced by market cycles and shifting narratives. In the longer term, patterns emerge—not precise, but directional. The system adapts, responds, and gradually reflects the forces acting upon it.

Those forces—structure, pressure, execution, and preservation—do not operate in isolation. Their influence shifts with context. In periods of expansion, access grows and infrastructure develops. In periods of stress, behavior adjusts and assumptions are tested. In consolidation, ineffective models are filtered out, and core properties are defended.

What matters is not identifying a dominant force, but understanding their interaction.

Participants experience this system differently depending on where they enter. For many, Bitcoin begins as an asset—something to acquire, hold, or trade. With deeper engagement, additional layers become visible: how transactions function, how custody is managed, and what tradeoffs exist between convenience and control.

Over time, exposure leads to reflection. Patterns become clearer. Reliability matters more than novelty. Questions of meaning surface alongside questions of use.

This progression—from exposure to understanding to reflection—is not universal, but it mirrors the system itself. Possibility becomes structure. Structure is tested. What survives is selected. What remains is examined for alignment with underlying principles.

Each new cycle introduces new participants, new conditions, and new constraints. Understanding does not come from a single moment, but from sustained observation.

This process is not passive. Engagement shapes perspective, and perspective evolves with experience. Bitcoin reflects the positions and priorities of those interacting with it.

This does not simplify the system. It makes it more coherent.

Multiple interpretations can coexist—not as contradictions, but as components of a larger whole. Movement between these perspectives becomes possible. The system is not fixed; it is continuously revealed through interaction.

Bitcoin does not converge into a single narrative because it evolves through tension, adaptation, and the accumulation of decisions. The relevant question is not what Bitcoin is, but what it is becoming.

That question cannot be answered definitively, but it can be approached by observing how forces interact, how patterns emerge, and how participants respond. The future is not predetermined. It is shaped through use.

Attention, then, must be directed not only at Bitcoin as it exists today, but at the processes that are shaping it—access pathways, behavioral shifts, execution constraints, and preservation efforts. These elements determine how the system develops.

At a closer level, the system can appear disordered. Each cycle introduces new narratives, new participants, and new interpretations. It may appear that Bitcoin is becoming something different. From a broader perspective, however, an unfamiliar pattern emerges.

Bitcoin is not changing its identity. It expresses different aspects of itself under different conditions.

It functions as an asset at times, while at other moments, as infrastructure, as a response to instability, or as a mechanism for preserving value. These are not contradictions but manifestations.

The perceived inconsistency lies not in the system, but in the expectation that it should conform to a single definition.

Once that expectation is removed, the system becomes easier to follow—not simpler, but more consistent in its behavior. What appears depends on the forces acting upon it.

Bitcoin is not revealed in isolation. It is revealed through participation—through decisions to build, allocate, connect, and preserve. Each action contributes incrementally. No central coordination is required.

The outcome cannot be predicted with precision—not because it is random, but because the interactions are not fully visible in advance. New participants enter, new constraints emerge, and new conditions test the system.

This does not show failure. It reflects openness. Bitcoin is not a closed system. It continues to form through use.

Its foundational properties remain stable—the protocol operates consistently, supply is fixed—but the meaning derived from those properties is context-dependent. It shifts with use, with conditions, and with interpretation.

Therefore, Bitcoin resists simple categorization. It operates across boundaries:

Between money and infrastructure.
Between individual control and institutional integration.

Between autonomy and coordination.

These boundaries are not resolved. They are navigated continuously by participants with different objectives.

This is not a flaw. It is the defining characteristic of the system.

Bitcoin does not require agreement to function. It requires participation. Through participation, its structure becomes visible incrementally.

No single perspective captures it completely. Each reveals a part of the system. Together, they map it imperfectly, but sufficiently to understand how it is being shaped.

That process does not end.

Conditions continue to change. Markets develop. Technology advances. New perspectives emerge.

Each shift introduces new dynamics and reveals aspects that were previously less visible.

The most accurate statement about Bitcoin is not a definition, but a recognition: it cannot be reduced to a single idea.

It must be observed across contexts.

Because what Bitcoin becomes will not be decided in a single moment.

It will be revealed.

About the Author

Jamil Hasan is the founder of Crypto Hipster Publications and host of the Crypto Hipster Podcast, a platform built around a single idea: where builders talk freedom, not price.

Across 580 conversations, he has focused on founders, entrepreneurs, and independent creators—the people actively building the digital economy from the ground up. His work does not center on market cycles or short-term narratives, but on understanding the deeper motivations, risks, and convictions behind those shaping new systems.

His approach is rooted in separating the signal from the noise. In an industry often driven by speculation and surface-level commentary, he has built a platform that prioritizes long-term thinking, first-principles discussion, and authentic perspectives.

Before fully committing to the digital asset space, Jamil built his career across financial services and data-driven initiatives, including prior work connected to blockchain, artificial intelligence, and client-facing strategy. That foundation informs his ability to translate complex technological change into clear, human insights.

After stepping away from his work during a period of intensive medical treatment, he returned with a sharper lens and a more defined mission: to elevate real builders, reject noise disguised as insight, and focus only on what endures.

Signals Through the Noise represents the next evolution of that mission.

Crypto Hipster Podcasts

A list of podcasts that form the basis of this book from Seasons 6-8 is presented below. All of them can be found at the Crypto Hipster Podcast station on Spotify, Apple Podcasts, Amazon, YouTube, Anchor, or wherever enjoy your favorite podcasts.

- How to Make Bitcoin Transactions Safe, Simple, and Accessible for Everyone, with Sung Choi @ Coinme

- Why Bitcoin Is Unbreakable in the Face of Black Swan Global Events, with Agne Linge @ WeFi

- Crypto Hipster Presents: Shooting from the Hip, Episode 1: The Meaning and Beauty of Bitcoin Maximalism, with Vlad Costea @ Bitcoin Takeover Podcast

- Bitcoin & Beyond: Pioneering Innovation in the Crypto Payments Industry with Mark Højgaard @ Coinify